# Astronauts

Giles Sparrow

**WORLD ALMANAC® LIBRARY**

Please visit our Web site at: www.garethstevens.com
For a free color catalog describing World Almanac® Library's list of high-quality books
and multimedia programs, call 1-800-848-2928 (USA) or 1-800-387-3178 (Canada).
World Almanac® Library's fax: (414) 332-3567.

Library of Congress Cataloging-in-Publication Data

Sparrow, Giles.
    Astronauts / by Giles Sparrow.
      p. cm. — (Secrets of the universe)
    Includes bibliographical references and index.
    ISBN-10: 0-8368-7275-4 — ISBN-13: 0-8368-7275-0 (lib. bdg.)
    ISBN-10: 0-8368-7282-7 — ISBN-13: 0-8368-7282-8 (softcover)
    1. Astronautics—Vocational guidance—Juvenile literature.  2. Manned space flight—
History—Juvenile literature.  3. Astronauts—Juvenile literature.  I. Title.  II. Series:
Sparrow, Giles. Secrets of the universe.  III. Series.
    TL793.S688   2006
    629.45—dc22                         2006009942

This North American edition first published in 2007 by
**World Almanac® Library**
A Member of the WRC Media Family of Companies
330 West Olive Street, Suite 100
Milwaukee, WI 53212 USA

This U.S. edition copyright © 2007 by World Almanac® Library. Original edition copyright
© 2006 by IMP. Produced by Amber Books Ltd., Bradley's Close, 74–77 White Lion Street, London
N1 9PF, U.K.

Amber Books project editor: James Bennett
Amber Books design: Richard Mason
Amber Books picture research: Terry Forshaw

World Almanac® Library editor: Carol Ryback
World Almanac® Library designer: Scott M. Krall
World Almanac® Library art direction: Tammy West
World Almanac® Library production: Jessica Morris and Robert Kraus

Picture acknowledgments: All photographs courtesy of NASA except for the following:
Getty Images: 21 (Mikhail Grachyev). All artworks courtesy of International Masters Publishers Ltd.

Printed in the United States of America

1 2 3 4 5 6 7 8 9 10 09 08 07 06

# CONTENTS

**Cover and title page:** Astronauts can perform necessary repairs and maintenance on satellites, the *International Space Station*, or the shuttle orbiter during extra vehicular activities (EVAs) while far above Earth. In March 1965, Russian cosmonaut Aleksei Leonov became the first person to walk in space.

## CHAPTER 1

# SPACE PIONEERS

In the years since the first human orbited Earth in 1961, more than four hundred people have flown in space. At times, spaceflight has become almost routine, but occasional emergencies and disasters in space are a stark reminder of the dangers involved. Although the first astronauts were military test pilots, many of today's astronauts are scientists. Current space travelers still take tremendous risks, but they all owe a great debt to the brave astronauts of the 1960s and 1970s—the true pioneers of the space age.

## The Space Race

In the early years, spaceflight was a competition between two rival superpowers—the United States and the former Soviet Union. With completely different political systems, the capitalist, democratic U.S. and the communist Soviet Union were uneasy allies during World War II (1939–1945), but became enemies soon afterward. A political "Cold War" followed. The space race developed because of the Cold War: each side rushed to gain superiority over the other by proving itself more accomplished in space.

In about 1955, the United States announced that it would put a satellite into Earth orbit during the upcoming International Geophysical Year, 1957–1958. The Soviets said they would also launch a satellite, and they beat the U.S. into space with the launch of *Sputnik* on October 4, 1957.

*Sputnik* shook the world and signaled the official beginning of the space race, an intense competition between the superpowers for space "firsts" and new records. It reached its finale in the U.S. Apollo Moon landings. Both countries poured money and expertise into their space programs. Without the political impetus behind the space race, however, space travel might have taken far longer to become a reality.

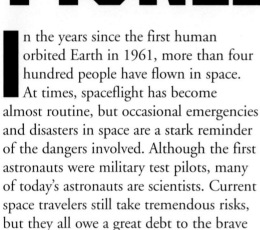

**John Glenn Jr. squeezes into the cramped confines of his Mercury capsule, *Friendship 7*. On February 20, 1962, he became the first American in orbit.**

## TO THE EDGE OF SPACE

Two years before the first man orbited Earth, a U.S. research aircraft took pilots to the edge of space for the first time. The X-15 was commissioned in 1955 as a prototype rocket-powered aircraft. It tested the behavior of aircraft at high altitudes and speeds and also tested the reactions of pilots to these conditions.

The X-15 was a stubby-winged aircraft launched by dropping from a B-52 bomber at about 50,000 feet (15,000 meters). Once clear of its carrier aircraft, the X-15's rocket engines kicked in, burning for about one hundred seconds and propelling the plane up to an altitude of about 55 miles (89 km). Here, the atmosphere is so thin that the X-15 could not fly like a regular plane— instead, the pilot had

### The first astronauts

Astronaut is the English name for a space traveler (from the Greek words for "star" and "sailor"). "Cosmonaut" is its Russian equivalent. (The French term is "spationaut," and Chinese astronauts are called "taikonauts.") The first person in space was a cosmonaut.

The early cosmonauts were selected from among the best pilots in the Soviet Air Force. In 1960, twenty-one pilots were chosen for space training. Soviet space program chief Sergei Korolev selected twenty-seven-year-old Yuri Gagarin and twenty-six-year-old Gherman Titov as possible pilots for the first manned spacecraft.

Gagarin was launched into space in the tiny *Vostok 1* capsule on April 12, 1961, entering an orbit that took him a maximum of 195 miles (315 kilometers) from Earth. *Vostok 1* made a single orbit of Earth in just 108 minutes, but the Soviet press did not wait for its safe return to Earth before announcing the news to the world. They almost regretted this when the flight nearly ended in disaster. When Gagarin's craft began the maneuvers that would return it to Earth, the reentry capsule failed to separate from the rest of the capsule. *Vostok 1* was left spinning out of control for several minutes before the cables connecting the two

parts burned through and the reentry capsule was released.

Gagarin's safe return and triumphant tour around the world was another blow to America's efforts to achieve space "firsts." The newly

**Mercury astronaut John Glenn Jr. became an all-American hero. He later entered politics and served as a U.S. Senator (1974–1998). In 1998, at age 77, Glenn returned to space aboard the space shuttle *Discovery*, becoming the oldest person to ever fly in space.**

to steer the X-15 using a series of rocket thrusters similar to those used on spacecraft. As the aircraft dropped from its maximum altitude, it became a high-speed glider (just like the modern space shuttle), and as the X-15 dropped back down through the atmosphere, the pilot flew it in to land using controls similar to a conventional aircraft. Several test pilots flew the X-15 in the early 1960s, including Neil Armstrong, who became better known as the first astronaut to walk on the Moon.

formed National Aeronautics and Space Agency (NASA) began selecting astronauts in 1959, recruiting candidates from the military's best pilots and engineers. It narrowed an initial list of five hundred to just seven men. When they completed their training and the Mercury spacecraft was ready, NASA promoted them in the media as the "Mercury Seven." One thing was missing, however: the new Atlas rockets needed to put Mercury into orbit. Anxious not to trail too far behind, on May 5, 1961, NASA launched Alan B. Shepard in the *Freedom 7* capsule using the underpowered Redstone rocket. Shepard's fifteen-minute flight reached a suborbital altitude of 115 miles (185 km). He became the first American to enter space.

Astronaut Gus Grissom (1926–1967) made the second suborbital Mercury flight aboard *Liberty Bell 7* in July 1961. By the end of the year, the Atlas rocket was finally ready. On February 20, 1962, *Friendship 7* was launched on a five-hour flight, carrying John Glenn Jr., the first American to orbit Earth. Three more Mercury flights followed: The seventh Mercury astronaut, Donald K. "Deke" Slayton, was grounded after doctors discovered a possible heart problem. Slayton finally flew in space in 1975.

While the Mercury missions were underway, the Soviet cosmonaut program continued. A total of six Vostok missions flew—Gherman Titov spent a whole day in space on *Vostok 2* in August 1961, while *Vostoks 3* and *4*, in August 1962, put two cosmonauts in orbit at the same time, just a few miles (km) apart.

*Vostoks 5* and *6* repeated the pattern of *Vostoks 3* and *4* in June 1963, but claimed another first. The pilot of *Vostok 6* was a woman, Valentina Tereshkova. An expert parachutist rather than a pilot, Tereshkova remained in orbit for three days, while her fellow cosmonaut, Valery Byskovsky, stayed in orbit for five days, setting a record for solo spaceflight that is still unbeaten.

# YURI GAGARIN

Yuri Gagarin (1934–1968) was born to parents who worked on a collective farm near Smolensk. He joined the Soviet Air Force after graduating from college and became a fighter pilot in 1957, before being selected for cosmonaut training in 1960. His first words from the capsule of *Vostok 1* were "I see Earth—it's so beautiful!" After the historic flight, Gagarin was paraded as a hero of the Soviet Union, but never flew in space again. He died in a crash during a test flight of a MiG jet fighter plane in March 1968.

# THE ELECTROMAGNETIC SPECTRUM

Light that we see is only a small part of the electromagnetic (EM) spectrum—the mostly invisible radiation, or energy, given off by stars. Electromagnetic radiation takes the form of different wavelengths of energy as it travels across the universe. All wavelengths of the EM spectrum move at the same speed: the speed of light—186,000 miles (300,000 km) per second.

   The visible part of the EM spectrum, in the middle, ranges from red light with longer wavelengths, to violet light, with shorter wavelengths. Beyond the visible violet light, the wavelengths become increasingly short, high-energy wavelengths that give off dangerous ionizing, or "hot" radiation, such as ultraviolet rays, X-rays, and gamma rays. Likewise, the wavelengths beyond red light become increasingly long, with lower energy levels, such as infrared (heat) waves, microwaves, radar waves, and radio waves.

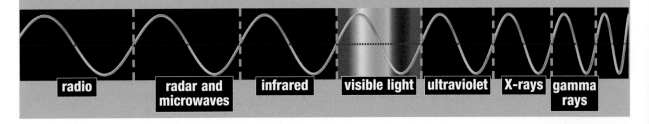

radio | radar and microwaves | infrared | visible light | ultraviolet | X-rays | gamma rays

## The Spacewalkers

The one-person Vostok and Mercury missions were followed by the more complex Soviet Voskhod and U.S. Gemini programs. These missions focused on practicing the techniques that would be needed for a mission to the Moon and for working in space. During these missions, astronauts left the protection of their spacecraft for the first time to "spacewalk" in orbit at the end of a long tether.

*Voskhod 1*, launched in October 1964, was a converted Vostok capsule with a three-man crew. In order to fit three people in, three small couches replaced the Vostok ejector seat. This meant that the three cosmonauts—Vladimir Komarov, Konstantin Feoktistov and Boris Yegorov—could no longer eject and parachute to the ground during landing. A "retrorocket" was attached to the base of the capsule, which would fire to slow down the capsule as it dropped toward Earth.

For *Voskhod 2*, the crew was reduced to two. This was necessary because one of the cosmonauts, Alexei Leonov, was dressed in a bulky space suit. When *Voskhod 2* reached orbit in March 1965, Leonov climbed out of a simple airlock and floated alone in space for twenty

minutes. Although he had no problems with the space walk itself, when Leonov tried to reenter the airlock, he discovered that his suit had expanded because of the pressure of air inside, and he could not fit back in. Leonov eventually had to release air from inside his suit in order to get back in the capsule.

The Americans caught up just a few months later when, in June 1965, astronaut Ed White made a twenty-one-minute space walk outside his *Gemini 4* capsule. Gemini was a series of ten two-man missions, which also practiced docking techniques in orbit, linking up with an unmanned Agena target spacecraft.

## The Apollo Program

By the mid-1960s, the U.S. was taking the lead in the space race, and preparations were underway for the Apollo Moon missions. These missions were the most complex spaceflights yet. They involved three linked spacecraft for different stages of the mission, and included some tricky maneuvers en route to the Moon. After several practice missions, the climax of the Apollo program occurred in 1969, when two of the three-person crew of *Apollo 11* climbed aboard the lunar module and

Cosmonaut Alexei Leonov during the first-ever space walk, part of the *Voskhod 2* mission of March 1965.

descended to the surface of the Moon, leaving astronaut Michael Collins in charge of the command module in lunar orbit.

The triumph of the Apollo program began in tragedy, when a fire killed three astronauts (including Ed White, America's first spacewalker) during a launch rehearsal of the *Apollo 1* capsule (*see page 42*). Several planned Apollo launches were cancelled or changed to unmanned test flights before crewed flights resumed with *Apollo 7* in October 1968, which was a simple test of the spacecraft in Earth orbit. *Apollo 8*, launched at Christmastime that year, took its astronauts—Frank Borman, William Anders, and James Lovell—on a rehearsal flight to the Moon, orbiting Earth's satellite ten times before returning. *Apollos 9* and *10* tested the separation of the lunar module in Earth orbit and in lunar orbit. *Apollo 10* astronauts Thomas Stafford and Eugene Cernan came within 9 miles (15 km) of the lunar surface before returning to a higher lunar orbit and heading back to Earth.

With all the different elements of Apollo tested, *Apollo 11* blasted off from Florida on

**Astronaut "Buzz" Aldrin inspects a package of experiments set up on the surface of the Moon during the *Apollo 11* mission of July 1969.**

July 16, 1969, carrying mission commander Neil Armstrong, Edwin "Buzz" Aldrin, and Michael Collins on their historic mission. Collins was the command module pilot. He remained in lunar orbit while Armstrong and Aldrin descended to the surface in the lunar module, *Eagle*. They landed in the Moon's Sea of Tranquility ("Tranquility Base") on July 20, 1969. A few hours later, Armstrong stepped onto the Moon's surface. The astronauts stayed on the Moon for only about twenty-one hours, but they had paved the way for later missions that remained on the lunar surface for up to three days.

## Further moon landings

Astronauts Pete Conrad and Alan Bean followed in *Apollo 12* in November 1969. They landed in the "Ocean of Storms" region of the Moon, close to the robot lander, *Surveyor 3*. Conrad and Bean

# OBSERVING ACROSS THE SPECTRUM

Only a small fraction of electromagnetic (EM) radiation from space reaches the surface of Earth. Although our planet's atmosphere absorbs most of the ultraviolet (UV) and some of the infrared (IR) and radio wavelengths, the visible portion of the EM spectrum makes it to the ground intact. We feel the IR radiation that penetrates the atmosphere as the Sun's heat on our bodies and other objects, while the UV rays that get through often produce skin damage, including tanning or sunburn. Still, the atmosphere also protects us from the more dangerous and damaging EM wavelengths, including X-rays and gamma rays.

We use the different wavelengths of the EM spectrum to explore space. Most ground-based telescopes scan the universe using visible light. For the clearest views, they are often located on mountaintops, where Earth's atmosphere is thinnest. On these mountain peaks, special IR telescopes also detect some of the IR radiation before the denser parts of our atmosphere block it. The best IR observing occurs from space-based telescopes, not only because of the lack of atmospheric blocking, but also because of the lack of ambient heat generated by Earth and by the IR telescope itself—which can distort images. The cold temperatures of space also require less refrigerant for cooling an orbiting IR telescope.

Earth-based radio telescopes, like the famous one in Arecibo, Puerto Rico, consist of huge metal dishes that collect long-wavelength radio waves from space. Smaller versions of radio telescopes, often built in movable groups called arrays, allow astronomers to combine many separate radio images into one larger image. Additionally, space-based radio telescopes collect and beam such data to Earth.

Space-based telescopes capable of studying the universe in different wavelengths became a reality in the decades after the launch of *Sputnik,* the world's first artificial satellite. While the famous *Hubble Space Telescope (HST)* collects images in visible light, it also carries equipment that scans the universe in IR—as does the *Spitzer Space Telescope (Spitzer).* Space-based UV instruments include the *Hopkins Ultraviolet Telescope,* used by space shuttle astronauts, the *Cosmic Hot Interstellar Plasma Spectrometer (CHIPS),* and the *Far Ultraviolet Spectroscopic Explorer (FUSE)* Mission. The *Wilkinson Microwave Anisotropy Probe (WMAP)* studies and maps the background microwave radiation of the universe. Space-based X-ray detectors include the *Rossi X-ray Timing Explorer* Mission, and the *XMM-Newton* and *Chandra* X-ray observatories, while the *High Energy Transient Explorer-2 (HETE-2)* Mission and *International Gamma-Ray Astrophysics Laboratory (INTEGRAL)* detect gamma-ray wavelengths. Telescopes dedicated to short-wavelength EM radiation are built to prevent these high-energy rays from simply passing right through them.

inspected this earlier lunar probe to see how it had survived in thirty months on the Moon.

*Apollo 13,* launched on April 11, 1970, never made it to the Moon. After an accident on the way, its crew, James Lovell, Fred Haise, and Jack Swigert, were lucky to return to Earth alive (*see pages 41–43*). After a delay in which the Apollo program was grounded while NASA made safety changes, *Apollo 14*'s Alan Shepard (the first American in space) and Edgar Mitchell landed in the "Fra Mauro" region of the Moon in February 1970.

The last three Apollo missions—*15, 16,* and *17*—carried lunar roving vehicles. The lunar

# NEIL ARMSTRONG

Neil Armstrong (b. 1930) was born in Wapakoneta, Ohio. After graduating with a degree in aerospace engineering, he joined the U.S. Navy in 1949 and served as a pilot during the Korean War. In 1960, he became a civilian test pilot for NASA, and made several flights in the X-15. He was selected as a NASA astronaut in 1962.

He first flew in space as commander of the *Gemini 8* mission, which he and fellow astronaut David Scott were lucky to survive (*see page 41*). *Apollo 11* was his last spaceflight. He left NASA in 1971, working in higher education and business until his retirement in 2002.

# THE SPEED OF LIGHT

All electromagnetic (EM) radiation travels through the vacuum of space at exactly the same speed—186,000 miles (300,000 km) per second. Most often, we call this the speed of light. (What we call "light" is the visible portion of the radiation of different wavelengths that makes up the EM spectrum.)

In his 1905 Special Theory of Relativity, Einstein's famous equation mathematically proved that nothing could travel faster than the speed of light. For this reason, we use the speed of light as a "constant"—a unit that never changes. One light-year is the distance light travels in one Earth year, which is roughly 6 trillion miles (10 trillion km). It is a convenient way of measuring the huge distances in space. In other words, a light-year measures distances, not time.

rovers allowed the astronauts to travel farther on the Moon's surface. The vehicles also increased the quantity of lunar rock samples astronauts could collect to bring back to Earth. Federal budget cuts led to the cancellation of three more planned Apollo missions, making *Apollo 17*, in December 1972, the last manned Moon mission.

## Space station pioneers

During the 1970s, the Cold War began to ease, and the space race came to an end. Both the United States and Soviet Union began to look at the long-term uses of space. Instead of following NASA to the Moon, the Soviets concentrated on an ambitious program of semipermanent space stations in orbit around Earth. They launched the first of these, named *Salyut 1*, in April 1971. It was a simple series of cylinders with solar panels attached to provide power in orbit. Crews flew to the station in Soyuz capsules, but the first mission turned back because of docking problems. The second crew carried out a successful mission in June 1971. They were killed, however, when their spacecraft, *Soyuz 11*, developed a leak on its return to Earth (*see pages 42-43*). The station itself was commanded to reenter Earth's atmosphere in October of that year.

Several more Salyut stations followed, with varied success. The Soviet military operated three of these early, top-secret Salyuts. Meanwhile, NASA launched its own space station, *Skylab*, in May 1973. *Skylab* had its own problems but hosted three successful missions in 1973 and 1974. *Skylab* crews carried out much useful science in the space station and also broke records for time spent in orbit. By this time, NASA began concentrating on building its reusable space shuttle, so plans for more U.S. space stations had to wait.

Meanwhile, Soviet space stations continued to improve. *Salyut 6* (1977–1981) and *Salyut 7* (1982–1986) were far more sophisticated than the earlier designs, which enabled cosmonauts to stay on board for much longer periods, eventually more than six months at a time. The gaps between missions grew shorter, and the "resident" cosmonauts could even receive visitors for short periods. In 1986, the Soviet Union launched *Mir* ("peace" in Russian), a completely new design of space station that could be expanded by the addition of extra modules. *Mir* was more or less permanently crewed between 1986 and 2000, and was the direct ancestor of today's *International Space Station*.

*Skylab* in orbit. The crumpled sheet in the center is an improvised sunshield built by the first crew after the original shielding and another solar panel ripped off during launch.

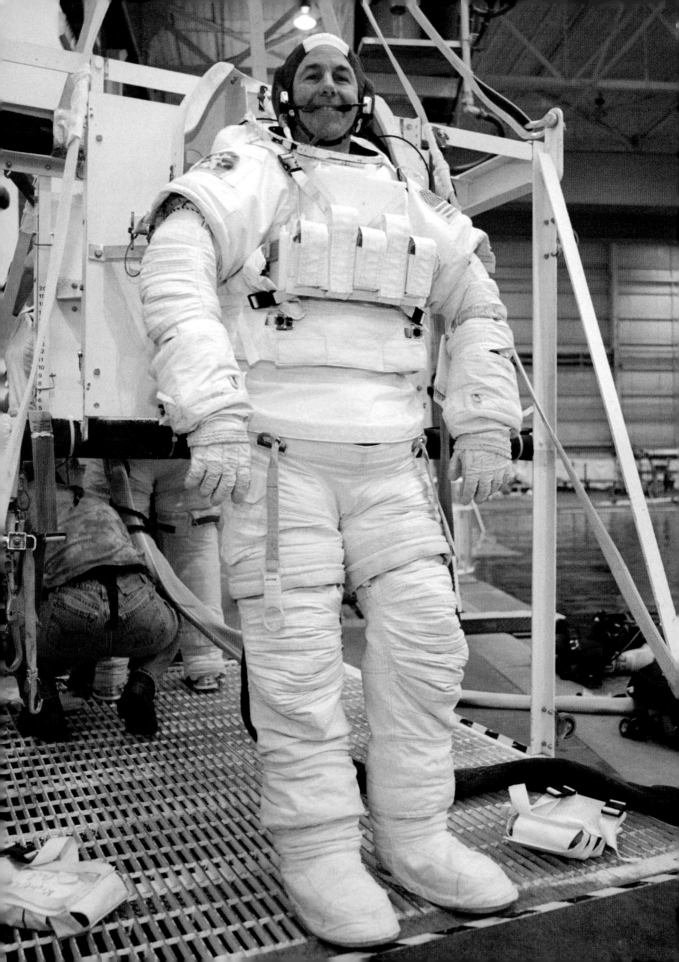

# BECOMING AN ASTRONAUT

## CHAPTER 2

In the early days of the space age, astronauts were mostly selected from a small, elite group of military test pilots. This began changing during the Apollo program as NASA slowly realized the advantages of sending scientists into space. As a member of the *Apollo 17* mission, geologist Harrison Schmidt became the first (and only) scientist to walk on the Moon.

## Astronaut selection

Today, the vast majority of astronauts go into space through the U.S. space program. NASA separates its astronauts into three different types—pilots, mission specialists, and payload specialists.

Pilot astronauts are the people who actually fly the space shuttle and are in overall charge of the mission. Each shuttle flight has two pilot astronauts on board—the mission commander and the pilot.

Pilot astronauts must have a degree in a scientific subject, and more than one thousand hours of experience flying jet aircraft, ideally as test pilots. Like all astronauts, their health and fitness levels must also meet strict standards. The pilots are the most vital people on any space mission, because if they cannot do their job, the entire crew may be endangered. Because of the flying experience needed, pilot astronauts usually come from the U.S. military. Anyone with suitable qualifications can, however, apply to the pilot astronaut program.

Mission specialists assist the pilot and commander with the general operation of the spacecraft—any duty that does not directly involve flying. Their main responsibility is for the mission's cargo, or "payload," carrying out experiments, or perhaps launching and repairing satellites. When necessary, mission specialists are the ones who leave the shuttle for space walks (known as extravehicular activities or EVAs). Mission specialists also help run the shuttle's onboard systems,

Astronaut Steve Robinson flew as a mission specialist on the space shuttle *Discovery*'s 2005 "Return to Flight" mission. Here, he wears a space suit adapted for underwater training.

monitor life support and supply levels, and plan the crew's daily activities.

Mission specialists must have a science degree and three years of professional experience in a related area. An advanced degree such as a master's or doctorate will often boost the chances of selection. The physical requirements are not quite as strict for these astronauts, but they must still be fit and healthy.

Payload specialists are astronauts selected for a mission by a third party, which is usually another space agency or a company that is flying a payload that requires special attention. Although payload specialists are not part of NASA's Astronaut Candidate Program, they are required to meet educational and physical standards and must participate in extensive training for their mission (*see below*).

Hundreds of qualified people apply to become pilot astronauts or mission specialists each year, so the final selections often depend on the applicants' performance in front of a series of

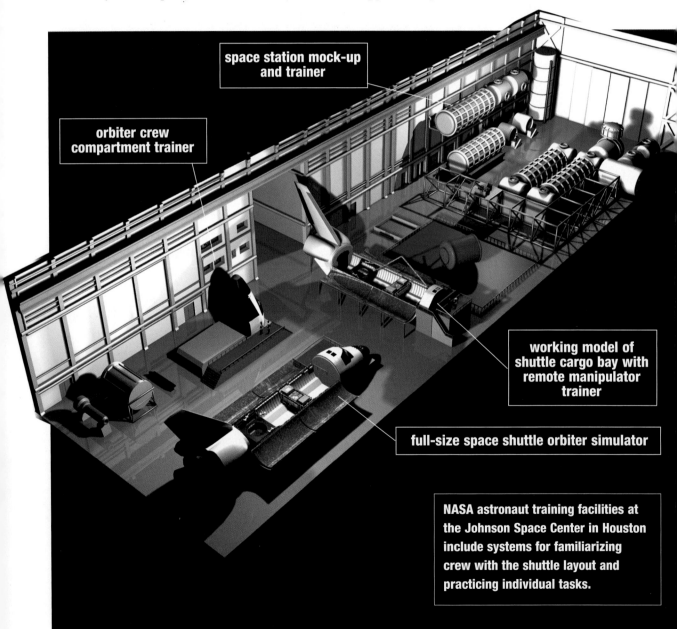

space station mock-up and trainer

orbiter crew compartment trainer

working model of shuttle cargo bay with remote manipulator trainer

full-size space shuttle orbiter simulator

NASA astronaut training facilities at the Johnson Space Center in Houston include systems for familiarizing crew with the shuttle layout and practicing individual tasks.

interview panels. NASA seeks people who can work well in a team, but who can also function independently when necessary. NASA also prefers candidates with useful skills and a good understanding of science and technology outside their own specialist areas. Successful military candidates serve with NASA for an agreed time period, while successful civilians join NASA's Astronaut Candidate Program as government employees and are expected to remain with NASA for at least five years.

Selection procedures for other space agencies differ. So far, China's astronauts have all been selected from the military, as have most Soviet and Russian cosmonauts. The European, Japanese, and Canadian Space Agencies also train a small number of astronauts. Generally, those selected are highly skilled professionals with military, science, or engineering backgrounds.

### Training for space

Once selected, NASA astronaut candidates begin a tough training program that lasts for one year at the Johnson Space Center (JSC) near Houston, Texas. The early weeks include refresher courses in many areas of science as well as basic survival training on land or at sea (in case of a forced landing). The astronauts must pass a swimming test and train as scuba divers in preparation for later stages of training that involve working underwater (*see page 21*). Other early tests include getting used to space suits, and coping with high- and low-pressure atmospheric conditions.

Flight experience is a vital part of training. Pilot astronauts are expected to maintain their skills by flying at least fifteen hours a month using NASA's own two-seater T-38 training jets. Mission specialists fly at least four hours a month throughout the training period as well. All astronauts must make several flights in an aircraft designed to simulate the weightless conditions of orbit (*see The "Vomit Comet" box*).

In conjunction with this basic training, the astronauts become familiar with the space shuttle

Up and down get turned around during zero-gravity training aboard the "Vomit Comet."

## THE "VOMIT COMET"

"Vomit Comet" is the nickname given by astronauts to the aircraft used in NASA's Reduced Gravity Research Program. Currently, NASA uses a converted McDonnell Douglas C-9 airliner, and also pays a commercial operator for flights on a converted Boeing 727. The plane climbs to a high altitude and then dives on a special flight path called a parabola, so that the aircraft and the astronauts inside it fall toward Earth at the same rate. In these conditions, the astronauts are in "free fall" inside the aircraft, and experience weightlessness for about twenty-five seconds of each sixty-five-second dive. The aircraft can complete up to forty of these dives in a typical flight. This allows the astronauts to get used to moving in weightless conditions—but can also give them their first experience of space sickness!

or, as NASA calls it, the Space Transportation System (STS). Training on various computer programs, including a number of mission simulators, follows.

Crew members first train on the single system trainers (SSTs), which familiarize them with each of the main systems on the shuttle orbiter (the part of the STS that actually makes it into space). The training process includes running through checklists that are used on the actual missions, identifying problems, and learning how to correct them.

After learning the individual systems, trainees move to the shuttle mission simulators (SMS), a series of computerized simulations that allow them to practice mission phases such as launch, satellite release (deployment), landing, and how to use the shuttle's robot arm—the remote manipulator system (RMS).

Once crews are assigned to a specific flight, they practice all phases of their mission on a full-sized working model of the shuttle orbiter. This simulator is somewhat like a giant virtual-reality video game, with screens at the windows that allow the

Pilot astronauts practice for launch in the cramped confines of a modified flight simulator. All shuttle crews wear pressure suits with helmets during takeoff, switching to more comfortable clothing once safely in orbit.

## EILEEN COLLINS

Eileen Collins (b. 1956) is the first female space shuttle commander. After studying mathematics, science, and economics, she joined the United States Air Force and became an instructor pilot on T-38 training jets, moving on to command C-141 Starlifter transportation aircraft.

From 1986 to 1989, she worked as an instructor at the United States Air Force Academy in Colorado Springs, Colorado, and then completed the astronaut training program in 1991.

Collins worked in support roles in engineering and Mission Control before becoming the first female shuttle pilot on *Discovery* in February 1995.

astronauts to see computer-generated views of the spacecraft and Earth below it, just as they would during a real mission. Other training simulators at the JSC include a full-scale model of the fuselage that helps trainees learn the layout and various day-to-day procedures. A replica of the crew section tilts vertically to help the astronauts get used to different flying positions.

The crew compartment of the orbiter has two main parts. The upper flight deck contains the main flight control systems, seats for the pilots and the rest of the crew, the operating system for the cargo bay, and the RMS. The lower mid-deck is used for eating, sleeping, and preparation for space walks. It also has an airlock door to the cargo bay.

Pilot astronauts also train in a moving model of the flight deck. This is similar to the flight simulators used by aircraft pilots, except that it has a greater range of movements and can tilt to point straight up to simulate launches. Upon assignment to a particular mission, pilots also begin training for landing in specially modified Gulfstream business jets. Training for a landing is crucial because the shuttle glides back to Earth without engines—so there is only one chance to land. Despite the lack of power, the shuttle travels at extremely high speeds and descends at a steep angle. In order to simulate this, the pilots of the training aircraft actually throw their engines into reverse as they "approach" the landing strip.

Astronauts also train on a mock-up of the shuttle cargo bay, complete with robot arm. Dummy satellites and a model of the *Hubble Space Telescope* (*HST*) help them practice satellite deployment and retrieval. During an actual mission, the pilot astronauts often perform these delicate tasks. Astronauts begin by practicing generalized tasks, but once assigned to a particular mission, they train together as a crew, rehearsing the specific tasks required of them during their flight. For the last few weeks of the training program, astronauts working in the SMS are linked to NASA's Mission Control Center, just as they will be on the mission itself.

Another major training facility at the JSC is the Neutral Buoyancy Laboratory—a giant water tank. Here, astronauts practice operating in a weightless environment and within the confines of a space suit (*see box page 21*).

## Training around the world

The *International Space Station* (*ISS*) created opportunities for astronauts from a number of countries, and training programs now account for this. The majority of the astronauts are still American or Russian, however. They are expected to speak both languages well in order to communicate with mission controllers at the JSC and at the Star City complex near Moscow, Russia. Pairs of space station astronauts are assigned to expeditions that last several months

After another flight as pilot on *Atlantis* in 1997, she commanded the 1999 *Columbia* mission that deployed the *Chandra X-ray Observatory* satellite into orbit. She also commanded the crucial "Return to Flight" mission of *Discovery* in July 2005, following the loss of *Columbia*, which broke up on reentry in February 2003.

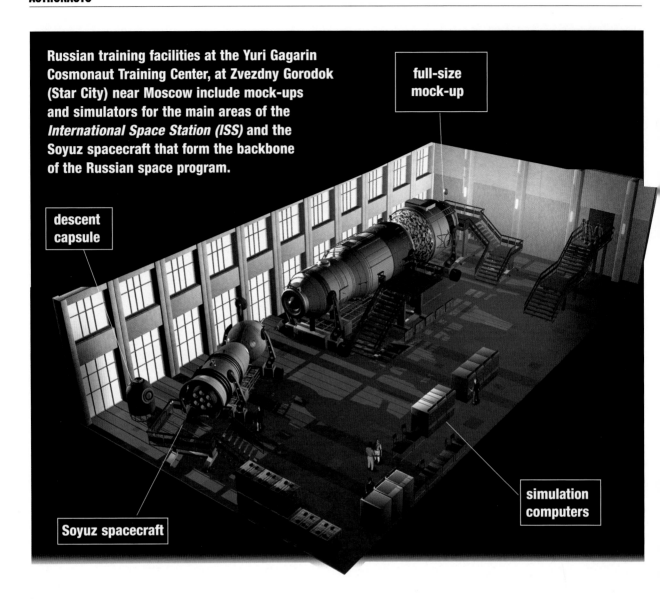

Russian training facilities at the Yuri Gagarin Cosmonaut Training Center, at Zvezdny Gorodok (Star City) near Moscow include mock-ups and simulators for the main areas of the *International Space Station (ISS)* and the Soyuz spacecraft that form the backbone of the Russian space program.

full-size mock-up

descent capsule

simulation computers

Soyuz spacecraft

at a time. They train at the JSC and at the Yuri Gagarin Cosmonaut Training Center, which is part of Star City.

Crew members who will travel to the *ISS* aboard the space shuttle participate in the shuttle training program and work with the other mission specialists for several months. Ever since the *Columbia* disaster (*see pages 44–45*) in 2003, *ISS* crews generally come and go in three-person Russian Soyuz spacecraft. These vehicles are far less complex than the shuttle, but crew members must still undergo intensive training.

As partners in the *ISS*, the European Space Agency (ESA) and Japan are also allowed to send

a number of astronauts to the station. These astronauts must be able to speak English and Russian as well as their native languages. Basic training that occurs at astronaut training centers in Germany and Japan, respectively, is designed to fit seamlessly into the joint U.S./Russian training program. ESA and Japanese astronauts then join their colleagues for more intensive training prior to the mission.

One final group of potential astronauts comes from space tourists—wealthy individuals who pay the Russian Space Agency up to $20 million for the chance to travel to the *ISS*. So far, three paying tourists have visited the station. The first

# FLOATING IN SPACE

Both the Johnson Space Center (JSC) and the cosmonaut training center at Star City use enormous water tanks to give astronauts more experience working in zero gravity. The Neutral Buoyancy Laboratory (NBL) at JSC contains a huge tank that holds 6.2 million gallons (23.4 million liters) of water, including a full-sized model of the *ISS* for astronauts to practice assembly and maintenance. When sealed into a modified space suit that is correctly weighted, each astronaut reaches "neutral buoyancy." He or she neither floats nor sinks in the water tank. The NBL isn't an exact simulation of conditions on a space walk. Despite that, it's still the closest thing to a weightless environment on Earth.

South African "space tourist" Mark Shuttleworth emerges from his *Soyuz TM-34* landing capsule after a successful return to Earth on May 5, 2002.

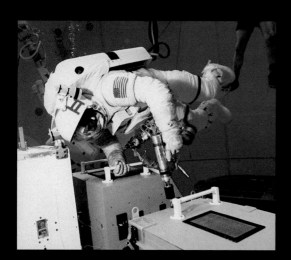

An astronaut in a specially adapted space suit practices zero-gravity maneuvers in the Neutral Buoyancy Laboratory at the JSC.

was sixty-year-old Dennis Tito, a U.S. businessman, who blasted into space on the *Soyuz TM-32* in April 2001.

Mark Shuttleworth, a twenty-nine-year-old South African millionaire who made his fortune with computers, flew to the *ISS* aboard *Soyuz TM-34* in late April 2002. And in October 2005, another sixty-year-old, U.S. scientist and entrepreneur Gregory Olsen, flew aboard the *Soyuz TMA-7*.

Space tourists usually join a Soyuz capsule taking a new replacement crew to the *ISS*, returning a few days later with the outgoing crew. Space tourists must participate in an intensive training program and pass various medical examinations before being allowed to fly.

## CHAPTER 3

# MANNED MISSIONS

**S**pace missions have come a long way since the early days of the space race. Although the early cosmonauts and astronauts were highly skilled pilots, they were sometimes reduced to almost the role of passengers on a fully automated flight. Today, they are expected to act as pilots, engineers, and scientists on spacecraft and space stations that are more complex than ever before. Once in orbit, they may conduct space walks, deploy or repair satellites, assemble space station components, or perform experiments in a weightless laboratory.

Despite all these changes, missions still begin in the same way, with a rocket-powered launch into orbit around Earth. Space launches are a spectacular sight, and a safe launch (and a safe return to Earth days or weeks later) involves hundreds of people on the ground as well as the skills of the astronaut pilots.

## Launch sites

Rocket launches require a huge array of special facilities and a large open area far from population centers. Rockets launch vertically, so special structures support the launch vehicle on the launchpad, releasing it when the countdown reaches zero and the rocket begins to move. The launch centers have enormous buildings for assembling the different sections of the rockets. The Launch Control Center (LCC), where the entire countdown procedure is monitored and controlled, sits at a safe distance from the launchpad.

The world's major sites for manned rocket launches are the Kennedy Space Center in Cape Canaveral, Florida, and the Baikonur Cosmodrome in Kazakhstan. Cape Canaveral, like many other launch sites (including the launch center in Kourou, French Guiana, for unmanned ESA missions), was built to take advantage of the speed boost offered by Earth's rotation. Close to the equator,

A Redstone rocket blasts free of its Florida launchpad on May 5, 1961. It carries the *Freedom 7* Mercury capsule containing Alan Shepard, the first American to reach space.

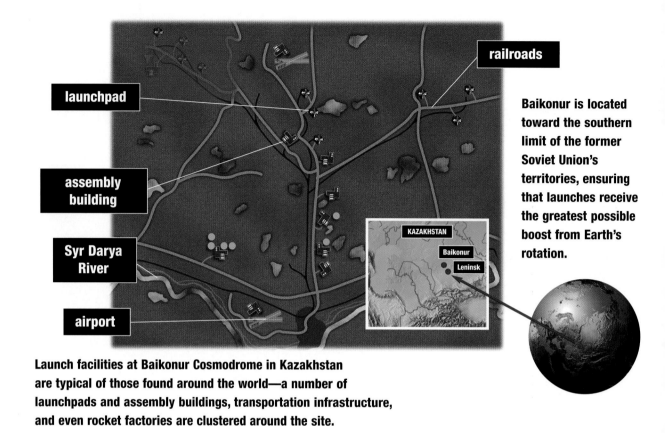

**railroads**

**launchpad**

**assembly building**

**Syr Darya River**

**airport**

KAZAKHSTAN

Baikonur

Leninsk

Baikonur is located toward the southern limit of the former Soviet Union's territories, ensuring that launches receive the greatest possible boost from Earth's rotation.

**Launch facilities at Baikonur Cosmodrome in Kazakhstan are typical of those found around the world—a number of launchpads and assembly buildings, transportation infrastructure, and even rocket factories are clustered around the site.**

any rocket launched in the direction of the Earth's spin will have its speed boosted by roughly 1,000 miles (1,600 km) per hour. Baikonur is situated south of the former Soviet Union, but not close to the equator because of the limits of Soviet territory when it was built. All these sites have large unpopulated areas or ocean in the direction of launch so that the lower stages of the vehicle can safely fall back to Earth once they exhaust their fuel and are jettisoned from the main rocket.

NASA launches the space shuttles from Complex 39 of the Kennedy Space Center. In the 1960s, this was the launch site for the Apollo program. The shuttle now uses many buildings originally built for handling the enormous Saturn V rockets that took humans to the Moon, including the fifty-two-story-tall Vehicle Assembly Building (VAB), one of the largest enclosed spaces in the world. The four main components of the shuttle system—the shuttle orbiter, liquid fuel tank, and two solid

rocket boosters—are assembled on a mobile launch platform and attached to towers that keep the entire vehicle upright. Once assembly is completed, a huge vehicle called the crawler transporter drags the launch platform and shuttle to the launchpad at about one mile (1.6 km) per hour. Once in place, the launch platform sits over a deep concrete hole called the blast pit. During launch, rocket exhaust fumes blow into this pit and escape along tunnels that release it a safe distance away from the spacecraft itself.

Once the shuttle is safely on the launchpad, arms swing out from the 347-foot (106-meter) "fixed service tower" to hold the orbiter in place. A larger "rotating service structure" (RSS) also swings into place, allowing access to the orbiter's crew cabin and cargo bay for final preparations and loading of supplies and payloads such as satellites. The ready shuttle often stands on the launchpad for several weeks before the final stages of countdown begin (*see page 25*).

# COUNTDOWN!

The countdown for a typical space shuttle launch is tightly scheduled and involves multiple checks on every aspect of the shuttle orbiter, fuel tank, and solid rocket boosters. Although many of these checks are automated, a launch still involves about 230 people in the Launch Control Center (LCC). It all begins with the "call to stations," about forty hours before launch. At this time, crew equipment is loaded, the shuttle's computers are activated, backup software is checked, and engineers make a final physical inspection of the shuttle.

• At "T" (target) minus 1 day, communications are established between the LCC and Mission Control in Houston. The air is pumped out of the orbiter and replaced with the inert gas nitrogen to reduce the risk of explosions during the fueling procedure.

• At T minus 6 hours, fuel is pumped into the huge external fuel tank. The liquid hydrogen fuel is stored at –423 °Fahrenheit (–253 °Celsius). The countdown is put on hold while the fuel tank is inspected for ice.

• At T minus 2 hours 30 minutes, the shuttle crew arrives at the launchpad and enters the orbiter. They are sealed in and make contact with the LCC launch and Mission Control.

• At T minus 9 minutes, the launch director in the LCC makes the announcement, "Go for launch." At this signal, the Ground Launch Sequencer (GLS), a computer that takes control of most of the countdown, is activated.

• At T minus 7 minutes 30 seconds, the orbiter access arm on the Fixed Service Structure retracts.

• At T minus 31 seconds, the signal to launch is sent to the shuttle computers. If necessary, the countdown can be stopped at this stage.

• At T minus 16 seconds, sprinklers flood the platform and permanent pad with water. This absorbs vibrations from the engine and prevents reflected noise from damaging the shuttle, but does not offer any significant fire protection.

• At T minus 10 seconds, valves open to allow fuel into the engine pumps.

• At T minus 6.6 seconds, the main shuttle engines ignite, gradually increasing to full power.

• At T minus 3 seconds, the solid rocket boosters ignite.

• At T minus 0, the explosive bolts that attach the shuttle to the launch platform fire, releasing the rocket. The shuttle slowly begins to move, gaining speed, and clearing the top of the Fixed Service Structure tower at T+7 seconds. At this point, Mission Control in Houston takes over control of the flight.

fixed service structure

service road

rotating service structure

space shuttle orbiter, fuel tank, and boosters

crawler transporter

exhaust pit

roadway for crawler transporter

Launch facilities vary for different countries and different vehicles. The shuttle's launchpad is probably the most complex, because the shuttle is a large and extremely complicated spacecraft. At the Baikonur Cosmodrome, where astronauts and cosmonauts are launched to the *ISS* by Soyuz rockets, the launchpad has a series of towers that rise out of the ground and lock into place around the vehicle.

Once a spacecraft is launched and safely on its way into space, Mission Control takes over from the LCC. In NASA's case, Mission Control is based at the Johnson Space Center in Houston, Texas. There, specialists work around the clock monitoring the status of spacecraft systems, communicating with the astronauts in orbit, and planning and coordinating mission activities, such as satellite releases and space walks.

## Working in space

Early spacecraft were so tiny that it was virtually impossible to do any serious work. Today, the smaller spacecraft are mostly used for short flights to and from space. Astronauts can only do real work on board the *ISS* and the space shuttle.

Shuttle missions use the spacecraft as an orbiting laboratory, as a delivery truck, or as a repair or construction vehicle. Delivery missions involve the deployment of satellites or space probes carried in the shuttle's cargo bay. The shuttle can carry two satellites with a total weight of up to 65,000 pounds (29,500 kilograms), and launch them from a spinning turntable into their new orbits. Space probes and satellites that need boosting to a much higher orbit are launched with a small rocket stage that fires when safely away from the shuttle and propels the vehicle into its planned orbit. Space probes launched this way include the *Galileo* mission to Jupiter and the *Magellan* probe to Venus.

With skillful piloting and operation, the shuttle can pluck satellites out of orbit and secure them in the shuttle's cargo bay for repair or to return the satellite back to Earth for maintenance. Probably the most spectacular

## GENE KRANZ

Eugene (Gene) Kranz (b. 1933) is probably the best known of NASA's mission controllers, because of his role in the *Apollo 13* mission (*see pages 41–43*). Kranz studied aeronautical engineering at St. Louis University, in St. Louis, Missouri, before joining the U.S. Air Force and flying jet fighters. He joined NASA in 1960, and was assistant flight director for the Mercury missions, moving on to act as flight director for the Gemini and Apollo missions. As flight director, he was in overall charge of Mission Control during these famous

early missions. He is pictured at left wearing one of the unique vests he wore while on duty (his wife made him a new one for each mission). Kranz was involved in the development of the software for operating the space shuttle and was promoted to director of mission operations in 1983. He retired from NASA in 1994.

Astronauts work to repair an Intelsat VI *F-3* communications satellite, snatched from orbit and successfully returned to space by the shuttle *Endeavour* in 1992.

The shuttle's mid-deck includes storage areas for a number of small experiments. For more ambitious science projects, a special laboratory module called *Spacelab* fits into the cargo hold. *Spacelab* also houses larger experiments, either inside the module (to take advantage of weightless conditions), or outside on special palettes (such as when an experiment needs exposure to the vacuum of space).

The main goal of space stations is to carry out scientific experiments in orbit over much longer periods. Stations can carry telescopes for astronomy, which is extremely useful because scientists have a much clearer view of space above Earth's atmosphere. Up there, they can also detect radiation that never reaches Earth's surface. And, various instruments study Earth itself from orbit.

Space station experiments often study the behavior of materials in zero gravity. Gravity distorts many high-tech materials. For instance, silicon chips "grown" from crystals on Earth differ from those grown in space: In weightless conditions, such silicon crystals grow quickly and perfectly.

On Earth, gravity distorts crystal structures during their formation (*right, upper*), and it is hard to create a precisely ordered material. In orbit, there are no forces interfering with crystal growth, and perfect structures (*right, lower*) are much more easily produced.

and successful shuttle repairs were the three servicing missions to the *Hubble Space Telescope*. Astronauts installed new optics and instruments that allowed the *HST* to operate well beyond its expected life. Those repairs increased our power to see the farthest objects in the universe.

The space shuttle also carried many *ISS* parts into orbit. *ISS* construction teams could then assemble those and any other *ISS* components sent into orbit earlier via unmanned rockets.

Although large-scale, space-based manufacturing is still in the future, scientists on Earth learn valuable lessons from studying how materials form in weightless conditions. Biological experiments often study how animals and plants adapt to zero gravity.

## Return to Earth

Apart from liftoff, reentry into the Earth's atmosphere is the most hazardous part of any spaceflight. The shuttle must turn around to face forward and use retrorockets to slow its speed to drop from orbit and fall toward Earth. The space shuttle spends most of its time in orbit facing backward. This position, with its tiled "belly" turned toward space, offers the crew and cargo maximum protection from solar radiation and space junk. As any spacecraft drops out of orbit into the upper atmosphere, it travels at extremely high speeds. Collisions with air molecules generate enormous amounts of heat that creates an envelope of burning gases, or plasma, around the spacecraft. The plasma can easily melt through weak points in the hull, causing a catastrophic explosion. This is why the space shuttle *Columbia* exploded.

Most spacecraft get rid of excess heat around them using an "ablative heat shield." Instead of just being heat-resistant, this type of shield is designed to burn and break away, carrying much of the heat with it. This is why most space capsules are cone-shaped with a heat shield covering their underside. The capsule simply turns so it descends base-first. As the heat shield breaks away, it falls harmlessly away from the unshielded parts of the spacecraft. Eventually, the combined effects of retrorockets and air resistance slow down the capsule enough for braking parachutes to be opened. These allow the spacecraft to slowly drift down to Earth.

Because the space shuttle was designed for reuse, its engineers developed an alternative to the heat shield. They decided to cover the vulnerable parts of the shuttle with ceramic tiles

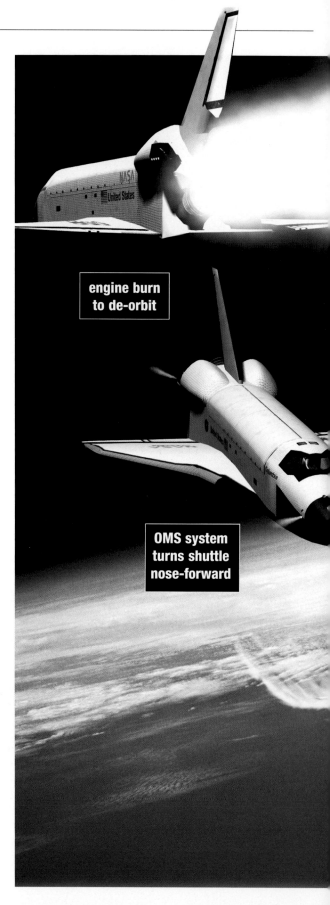

engine burn
to de-orbit

OMS system
turns shuttle
nose-forward

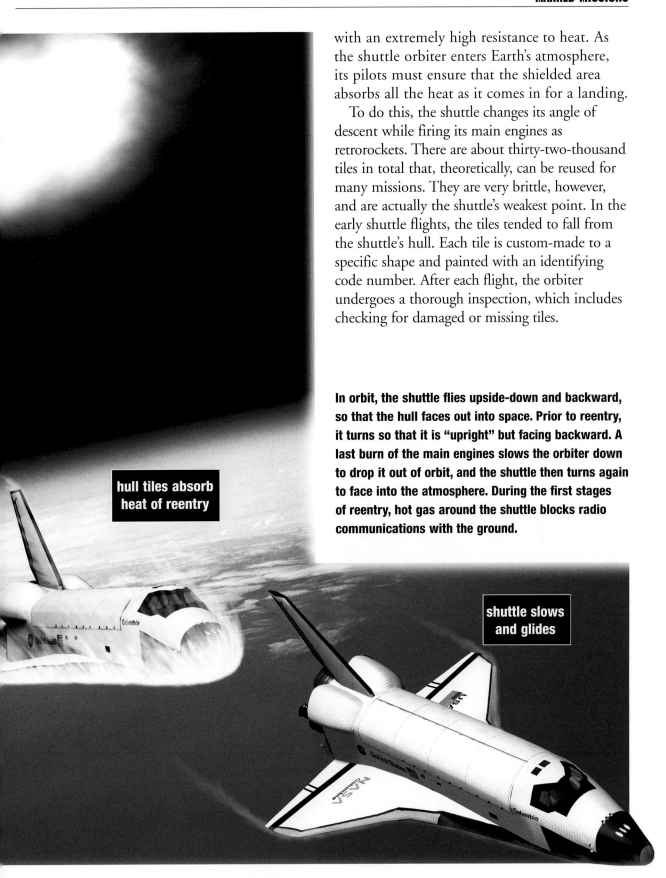

with an extremely high resistance to heat. As the shuttle orbiter enters Earth's atmosphere, its pilots must ensure that the shielded area absorbs all the heat as it comes in for a landing.

To do this, the shuttle changes its angle of descent while firing its main engines as retrorockets. There are about thirty-two-thousand tiles in total that, theoretically, can be reused for many missions. They are very brittle, however, and are actually the shuttle's weakest point. In the early shuttle flights, the tiles tended to fall from the shuttle's hull. Each tile is custom-made to a specific shape and painted with an identifying code number. After each flight, the orbiter undergoes a thorough inspection, which includes checking for damaged or missing tiles.

In orbit, the shuttle flies upside-down and backward, so that the hull faces out into space. Prior to reentry, it turns so that it is "upright" but facing backward. A last burn of the main engines slows the orbiter down to drop it out of orbit, and the shuttle then turns again to face into the atmosphere. During the first stages of reentry, hot gas around the shuttle blocks radio communications with the ground.

hull tiles absorb heat of reentry

shuttle slows and glides

# WORKING IN SPACE

## CHAPTER 4

**W**hen a spacecraft reaches orbit, it stops accelerating and enters a state of constant free fall. Everything on board becomes "weightless" (*see box, page 32*), and many astronauts experience "inversion illusion"—the fancy name for space sickness. Symptoms can include nausea and headaches. Most recover within a day or so, and medications often relieve symptoms. Training flights that simulate weightlessness (*see page 17*) help them prepare. Space sickness occurs when an astronaut's vision and body position do not seem to make sense to his or her brain, mostly because there is no "up" or "down" in space. Without the tug of gravity on their bodies, orbiting astronauts also tend to develop puffy faces as their blood drifts upward.

## Everyday life in zero gravity

Living and working in weightless conditions takes time to get used to, but it has advantages. For example, the cramped conditions inside spacecraft can be made more spacious by mounting the various work stations and day-to-day devices needed by the astronauts on the "floor" and "ceiling" as well as on the "walls."

In the early days of spaceflight, scientists worried about how astronauts would eat and drink. At first, many feared that human digestion might not function without gravity. This did not turn out to be a problem, but there was still concern about the potential mess that food could cause. Crumbs might float everywhere, and without gravity, water would pull itself into tiny droplets that could float around to potentially damage electrical circuits. The early Mercury missions experimented with various ways of packaging food, including cubes that could be chewed up in the mouth, dried foods that were swallowed with a gulp of water from a carton, and pastes that were squeezed out of a tube.

The *International Space Station* hangs in orbit above Earth. Its huge solar panels extend to 240 feet (73 m)—wider than a football field.

# WEIGHTLESS IN ORBIT

Earth's gravitational pull extends for millions of miles into space. A spacecraft in orbit a few hundred miles up is experiencing virtually the same amount of gravitational pull as an object on the surface of Earth. So why do astronauts feel weightless?

"Zero gravity" is actually a state of continuous free fall. Free fall is caused by the basic laws of motion discovered by English physicist Isaac Newton (1642–1726) in the 1600s. Newton showed that all objects stay still or (if already moving) keep moving in a straight line unless a force acts on them to change their movement. Such forces can include gravity or the thrust from a rocket.

A stable orbit is simply a path through space in which the spacecraft's tendency to fly off into space is balanced by the pull of Earth's gravity. Because the forces toward and away from Earth are perfectly balanced, the spacecraft simply circles

Earth until some force is applied to change the craft's speed. Since the astronauts are carried on the same path through space as their spacecraft, they experience the same apparent disappearance of gravity.

The temptation to fool around in weightless conditions is almost irresistible. Here—on board the U.S. *Skylab* space station—one astronaut performs the seemingly impossible feat of balancing another, head-down, on his fingertip.

These early food experiments were designed to meet an astronaut's daily nutrition needs, but they were tasteless and unsatisfying. For the Gemini and Apollo missions, astronauts carried dried foods in plastic packaging. Water was pumped into the pack, and the resulting paste squeezed out the other end.

Astronauts on the space shuttle and *ISS* benefit from a much wider range of food, in a form that most of us would recognize. These spacecraft carry a variety of prepackaged "ready meals,"

have a kitchen area with hot and cold water dispensers for rehydrating some of the packages, and an oven for heating them up. Once meals are cooked, the dishes clip onto special trays that can attach to an astronaut's lap, to the walls of the space shuttle, or to the mess table on the *ISS*.

Sleep is another basic requirement for all astronauts that often causes problems. For one thing, it is easy to lose track of the cycle of day and night on a spacecraft that sees a new sunrise every ninety minutes. NASA does its best to operate a twenty-four-hour day on the shuttle and the *ISS*, but astronauts tend to lose a couple of hours sleep each night, and gradually drift out of sync with the time on Earth. Sunlight streaming through the orbiter windows is so intense that it's impossible to use artificial lighting to create day and night on the shuttle. This means the only way astronauts can fall asleep is by wearing eye masks just like the ones distributed on overnight airline flights.

Zero-gravity conditions also interfere with sleep. Without constant attention, astronauts find themselves floating around the cabin while trying to sleep—or at least find that their arms drift around. For this reason, they sleep in special sleeping bags that anchor to the shuttle walls or attach to a special bunk system. Some also use arm straps.

## Life support systems

Keeping astronauts safe and comfortable in space requires a huge amount of technology. They need air, water, and food, as well as a way to dispose of wastes. The spacecraft must also protect the crew from extremes of heat and cold and the vacuum and radiation of space.

Air supplies on today's spacecraft use a mixture of 80 percent oxygen, 20 percent nitrogen—the same ratios found in Earth's atmosphere. Astronauts breathe at normal

**In space, even the most natural processes become very complicated. *Skylab*'s toilet involved advanced technology that simulated the natural assistance given by gravity.**

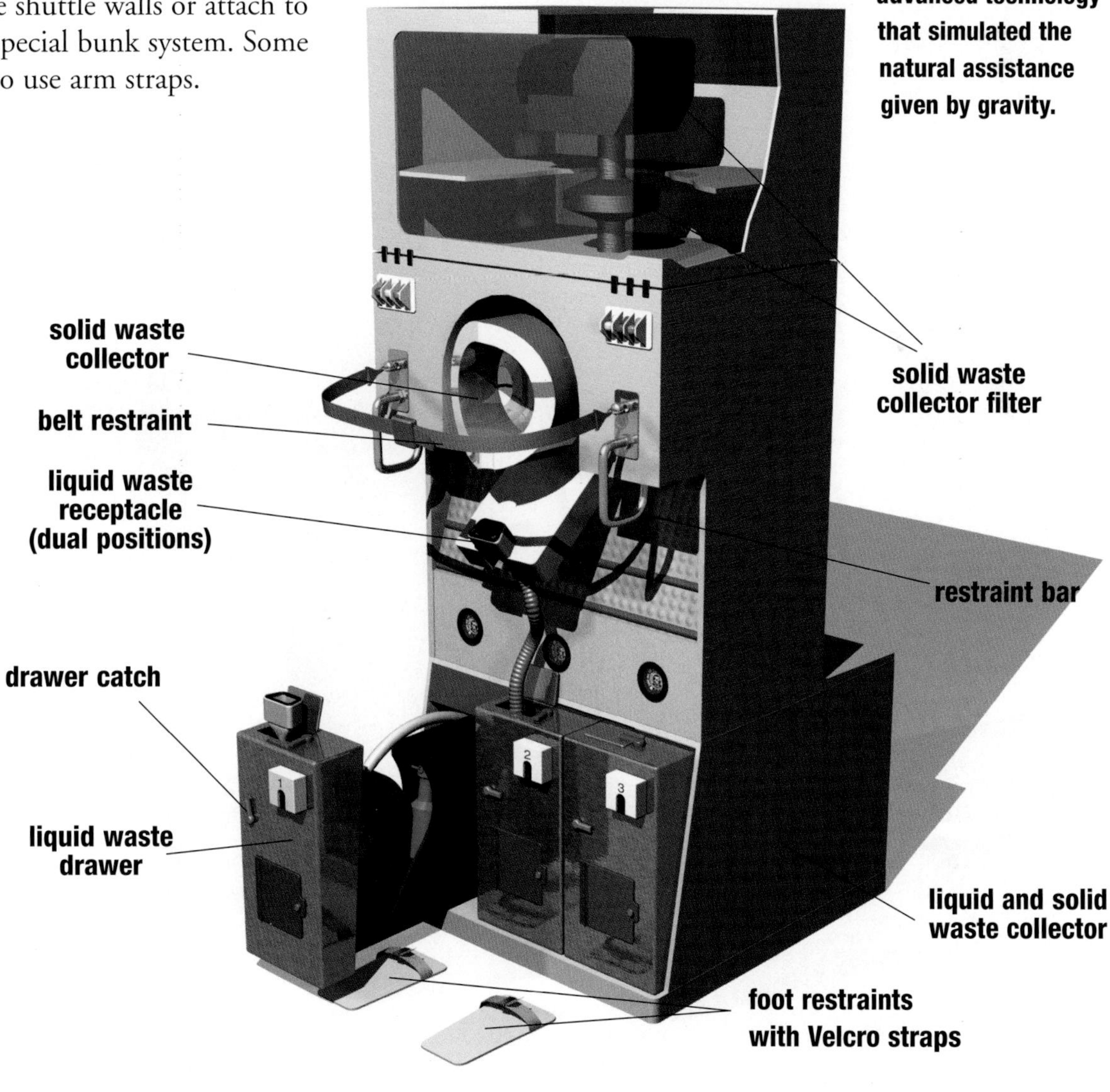

solid waste collector

belt restraint

liquid waste receptacle (dual positions)

drawer catch

liquid waste drawer

solid waste collector filter

restraint bar

liquid and solid waste collector

foot restraints with Velcro straps

# SPACE HYGIENE

Keeping clean is difficult in the cramped confines of a spacecraft, and some ingenious devices help astronauts with their hygiene. Free-floating water from a shower would be dangerous, wasteful, and would simply cling to an astronaut's body. Instead, astronauts cleanse themselves with damp sponges. They wash their hair using special shampoo that wipes off with a towel with no need for rinsing. It's also important not to let loose hairs escape from the towel and float free throughout the cabin. When astronauts need to use the bathroom, the zero-gravity environment complicates matters. Space toilets on the shuttle and the *ISS* use a vacuum and suction to imitate the effect of gravity. For urination, each astronaut uses a personal device attached to a suction tube, while for bowel movements, astronauts strap themselves onto a sit-down toilet. A filtering system keeps the air fresh.

**The shower used on *Skylab* incorporated a waterproof shower curtain to prevent leaks. After the astronaut had washed, the waste water was sucked out with a vacuum into a disposable bag.**

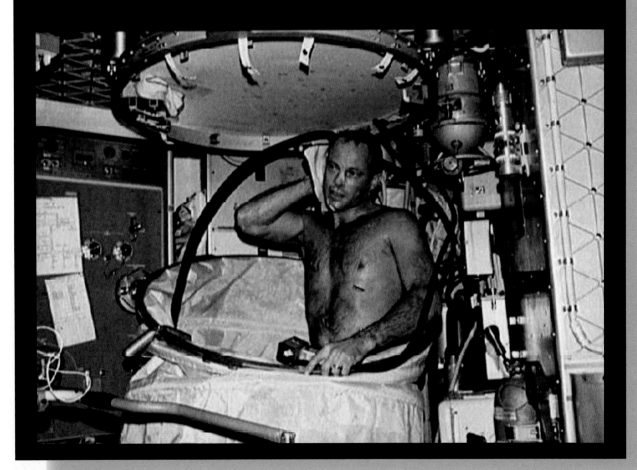

Earth atmospheric pressure. Gases are stored in pressurized tanks on both the space shuttle and the *ISS*. The tanks on the *ISS* are refreshed with new supplies from Earth as needed.

As the crew breathes in oxygen, they breathe out carbon dioxide gas as a waste product. If large concentrations of carbon dioxide build up inside the spacecraft, they can become toxic, so all spacecraft use devices that "scrub" the air. These are usually filters impregnated with lithium hydroxide, a chemical that absorbs the waste gas. They also use activated charcoal (the same kind used in household air fresheners) to remove odors.

In orbit around Earth, temperatures can range from −185 to 250 °F (−120 to 121 °C) in a matter of minutes. Spacecraft are well insulated to protect their interiors from these extremes of temperature. Spacecraft also have a "cryogenic" cooling system that works somewhat like a giant refrigerator. Low-temperature fluids circulate in tubes throughout the spacecraft to cool it.

Radiation is more of a problem, and there is still no adequate way to shield spacecraft from many of the high-energy particles flying around in space. Many astronauts have reported seeing random flashes, called the "Cerenkov effect," caused by cosmic rays. On Earth, the atmosphere absorbs such rays. In space, however, these rays pass through astronauts' eyes without causing problems. Fortunately, various medicines counteract the effects of other radiation, and astronauts have not yet stayed in space long enough to suffer much radiation damage.

## Staying in shape

Long periods in space have various effects on an astronaut's body. When we exercise on Earth's surface, gravity constantly pulls on our bodies, and part of our energy goes into keeping us from collapsing on the ground. It's easy to exercise by running,

doing push-ups, or using weights or other machines on Earth. In space—without the constant pull of gravity—many forms of exercise are impossible. Muscles weaken, and bones lose strength-giving calcium. Body fluids, such as blood, no longer "pool" in the legs but stay around the internal organs, and the body, "thinking" it has an ample supply of oxygen-carrying red blood cells (RBCs), reacts by reducing the amount of new RBCs produced.

Many of these problems are not felt until astronauts return to Earth. They may find that their weakened muscles can no longer support them against Earth's gravity, or that the lack of oxygen to the brain as blood sinks back down into their legs leaves them feeling faint or dizzy. So far, no astronaut has suffered from seriously weakened bone structure on returning to Earth. The problem must be solved before we launch a mission to Mars, however, which might involve years of space travel.

The best cure for muscle weakness is exercise, and most space stations have carried a variety of exercise machines to help the astronauts keep their muscles toned. Some cosmonauts have worn spring-loaded suits that force their muscles to work constantly.

## Space suits

A space suit is really a one-person spaceship. When an astronaut performs a space walk, or extravehicular activity (EVA), the space suit must provide all life-support functions. In the early days of spaceflight, astronauts often wore their space suits, complete with helmets, throughout their flight, in case an accident caused a leak or malfunction in the spacecraft's air supply. As space missions increased

**Space foods come in packaging designed for safe rehydration and consumption in weightlessness.**

# SHUTTLE EMU

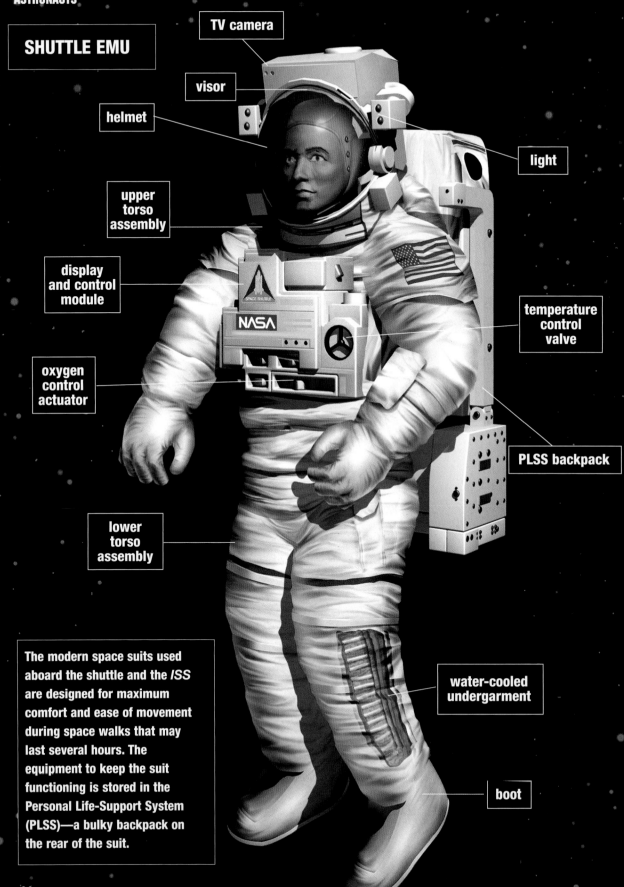

TV camera

visor

helmet

light

upper torso assembly

display and control module

temperature control valve

oxygen control actuator

PLSS backpack

lower torso assembly

water-cooled undergarment

boot

The modern space suits used aboard the shuttle and the *ISS* are designed for maximum comfort and ease of movement during space walks that may last several hours. The equipment to keep the suit functioning is stored in the Personal Life-Support System (PLSS)—a bulky backpack on the rear of the suit.

## LAYERS OF A SPACE SUIT

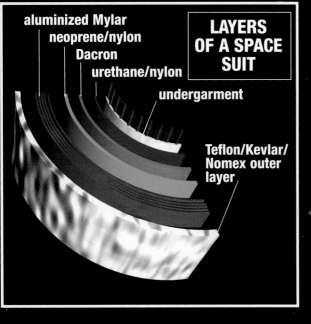

aluminized Mylar
neoprene/nylon
Dacron
urethane/nylon
undergarment

Teflon/Kevlar/Nomex outer layer

## PLSS BACKPACK

radio antenna

cooling unit

radio

primary oxygen

water

17-volt battery

secondary oxygen tanks

SAFER emergency propulsion unit

# VALERI POLYAKOV

Russian astronaut Valeri Polyakov (b. 1942) holds the record for the longest time spent in space. He remained aboard the *Mir* space station for 438 days in 1994–1995. Polyakov was first selected as a cosmonaut in 1972, but did not fly until 1988, in a mission lasting 240 days. As a specialist in space medicine, he used himself as a "guinea pig" for experiments during both of his long-duration missions. On his return to Earth in 1995, he retired as a cosmonaut but has continued to work in astronautical medicine.

Cosmonaut Valeri Polyakov, on board *Mir*, watches the approach of the shuttle *Discovery* in February 1995, partway through his record-breaking 438-day spaceflight.

in length, this was no longer practical. Today, most astronauts wear more comfortable clothes on board the spacecraft, and only don a space suit when leaving the *ISS* or shuttle for a space walk. Shuttle astronauts still wear orange pressure suits, similar to those worn by fighter pilots, during launch and landing, in case the shuttle's cabin loses pressure. These suits are designed to maintain pressure around the body at close to normal levels, even when the surrounding area is at very low pressures or absent completely. The suits are needed because the boiling point of liquids drops at low pressures, and above about 13 miles (21 km), human blood will boil from body heat alone.

Early space suits were tailor-made for individual astronauts, but those on the space shuttle and *ISS* are designed for general use. A variety of different elements can be linked together to suit different body shapes and sizes. The inner layer consists of a bodysuit cooled by liquid running through more than 300 feet (90 m) of tubing. The outer sections of the suit come in two halves. The lower section is flexible, with built-in boots and four layers: an airtight membrane; a layer of Dacron that tightens around the astronaut's lower

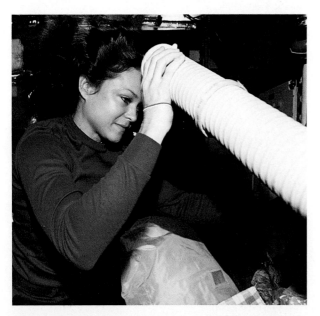

**An astronaut uses a vacuum hose to suck up stray hairs after washing her hair on board the shuttle.**

body; a thermal insulation layer; and a tough outer protective layer. The upper half has a solid torso with a built-in helmet, flexible arms with fitted gloves, and a backpack that contains oxygen supplies, water, a radio for communication, and fans to keep air moving around inside the suit. The two halves fit together with a connecting ring at the waist. Modern Russian space suits are designed in one piece, with a built-in backpack that slides aside to reveal a hatch that allows the cosmonaut into the suit.

Despite all these protective layers, astronauts still face the danger of overheating or freezing during space walks. On shuttle missions, space walks are scheduled so that the time astronauts spend in the harsh environment of space is balanced between periods of hot daylight and cold nighttime. The shuttle can even be rolled over in its orbit to ensure the best conditions for the spacewalkers. Space walks from the *ISS* must sometimes occur during cold periods, and the astronauts occasionally need to work in the shadow of large structures, so *ISS* space suits have been adapted to make them warmer. Astronauts can switch off the ventilation fans and use heating elements in their gloves to keep their fingers warm. They also have helmets fitted with floodlights and spotlights to help them work in the dark. Other changes to the *ISS* space suits allow the astronauts to service them between EVAs, which means that the suits are only returned to Earth for complete reservicing after twenty-five space walks.

## Space walk equipment

One benefit of the cramped conditions inside a spacecraft is that there is usually something to push or pull against. In weightless conditions, moving around outside a spacecraft can be extremely difficult. Outside the spacecraft and in a vacuum, astronauts cannot even "swim" through the air. If a spacewalking astronaut drifts too far from the ship, he or she could simply float off. Forever out of reach, he or she would find it impossible to get back.

Payload commander Steven L. Smith works on repairing the *Hubble Space Telescope* during a servicing mission aboard the space shuttle *Discovery* in December 1999.

Early spacewalkers were tethered to their spacecraft for safety, but they still had great problems moving around. Ed White, the first American to walk in space, carried a small "gun" with a gas canister that produced jets of gas when he pulled the trigger. The gun worked just like a miniature rocket—firing in one direction pushed the astronaut the opposite way. The major problem was that the gas canister was soon exhausted, and a larger gun would be too bulky to use on longer space walks.

For this reason, NASA and the Soviet Union both developed self-propelled backpack units during the 1980s. NASA experimented with a "jetpack" during later Gemini missions, but astronauts finally mastered the technology aboard the *Skylab* space station in the early 1970s, when they had room to experiment in the forgiving internal environment of the space station. The resulting device was called the manned maneuvering unit (MMU).

The MMU is a bulky pack that carries about 25 pounds (11 kilograms) of pressurized nitrogen gas in two cylinders, which is enough for more than an hour of spacewalking. The pack has two fold-down armrests fitted with controls designed for use through bulky space suit gloves. It can fire jets of gas at different pressures through twenty-four separate thruster nozzles that push the astronaut in almost any direction. Mission specialist astronauts train using the MMU in a special simulator—a replica MMU mounted on a crane that is operated by signals from the MMU's controls. Even with the MMU, walking in space is a slow and difficult business—the unit's normal speed is just 0.6 miles (1 km) per hour.

Astronauts on the *ISS* usually make tethered space walks, climbing around using handholds built into the station's structure. They can also use a simplified version of the MMU, known as SAFER (Simplified Aid for EVA Rescue), which can propel them back to the station if they accidentally come loose.

# DANGERS OF SPACE EXPLORATION

## CHAPTER 5

Despite the many technical advances in the decades since Yuri Gagarin first orbited Earth, spaceflight remains a risky venture. Every flight combines experimental technology, highly explosive fuels, and the ultimate hostile environment. With so much potential for problems to develop, it's not surprising that tragedies have occurred. Eighteen astronauts have died during spaceflights, and three more perished during ground training. There have also been several narrow escapes.

Perhaps surprisingly, both the U.S. and Soviet space programs survived their first few years without any catastrophes, although there were some close shaves: Gus Grissom nearly drowned on his 1961 suborbital Mercury flight when his capsule flooded during splashdown; Alexei Leonov had trouble reentering *Voskhod 2* after his historic spacewalk (*see page 9*); and Neil Armstrong and David Scott hit trouble when their *Gemini 8* capsule began spinning uncontrollably because of a misfiring thruster.

This record of narrow escapes ended tragically in January 1967 with the fire on board *Apollo 1* during a countdown test (*see box, page 42*). The Soviet Union also lost its first cosmonaut while testing a potential lunar spacecraft. Vladimir Komarov, pilot of *Soyuz 1* (the ancestor of the Soyuz modules still in use today) struggled to bring his spacecraft through reentry when problems developed with its steering rockets. Tragically, his descent parachute then tangled, and his capsule plummeted at great speed into the ground in central Russia.

The incredible survival story of the *Apollo 13* astronauts—the story of the mishap became a highly successful 1995 movie—is probably the most famous space accident. It occurred in April 1970, when public interest in spaceflight was near its height. *Apollo 13*'s difficulties began two days into its flight, when it

A solid rocket booster shoots free of the devastating explosion that destroyed the *Challenger* shuttle orbiter in January 1986.

# THE APOLLO 1 FIRE

Gus Grissom, Ed White, and Roger Chaffee, the first U.S. astronauts to die in the space program, formed the crew of *Apollo 1.* They died in an unfueled rocket during a routine launch rehearsal on the ground. Design errors in the capsule caused a spark between two poorly insulated wires that started a fire underneath Grissom's seat. Flames spread rapidly in the pure-oxygen atmosphere then used on U.S. spacecraft. It took rescuers ninety seconds to open the escape hatch. All three astronauts had died almost instantly in the intense blaze. An inquiry criticized the standards of workmanship throughout the Apollo capsule. NASA made many changes, including the use of a less-flammable nitrogen/oxygen mix for the air supply, before any Apollo missions flew again.

**escape hatch jammed during rescue attempt**

**atmosphere was pure oxygen**

**fire began after short-circuit in wiring under couches**

**flammable material in couches created toxic smoke**

**Above, right: Diagram of the *Apollo 1* command module, within which three astronauts perished during a launch rehearsal in January 1967.**

**Above (left to right): The three victims of the *Apollo 1* tragedy—Roger Chaffee, Virgil "Gus" Grissom (America's second man in space), and Ed White (America's first man to walk in space).**

was already too late to turn back. A routine test triggered an electrical short-circuit and caused an explosion that left the command module powerless and leaking oxygen. To survive, James Lovell, Jack Swigert, and Fred Haise climbed into the cramped lunar module (built for two people) and relied on its life-support systems while they made a hazardous journey around the Moon. They had to improvise a new way of using the lunar module's retrorockets to blast out of lunar orbit and back toward Earth. Eventually, after nearly six days in space, they splashed down successfully in the Pacific Ocean.

Just over a year later, the Soviet space program suffered another tragedy in what should have been a moment of triumph. The three-man crew of *Soyuz 11* had just set a new record for space endurance. After successfully docking with the world's first space station, *Salyut 1*, they spent twenty-four days on board. As they returned to Earth, and the reentry capsule separated from the rest of their spacecraft, a valve accidentally opened, and air leaked into space.

Left to right: Cosmonauts Georgi Dobrovolsky, Viktor Patseyev, and Vladislav Volkov set a record aboard *Salyut 1*, but died on the way home while aboard *Soyuz 11*.

By that stage in the Soviet space program, crews no longer wore space suits all the time. Georgi Dobrovolsky, Viktor Patseyev, and Vladislav Volkov rapidly suffocated. Everything appeared normal to the ground controllers and support crew watching the descent until they opened the capsule and discovered the tragic ending to an otherwise successful mission.

Soviet astronauts also had a couple of narrow escapes in the next decade, both caused by problems with their rockets. In 1974, the crew of *Soyuz 18*, en route to *Salyut 4*, was almost killed when the second and third stages of their rocket failed to separate, and it went out of control. An emergency system blasted the spacecraft clear, and the crew deployed a parachute and drifted back to the ground. In 1983, the crew of *Soyuz T10* faced a similar narrow escape when their rocket caught fire and exploded on the launchpad.

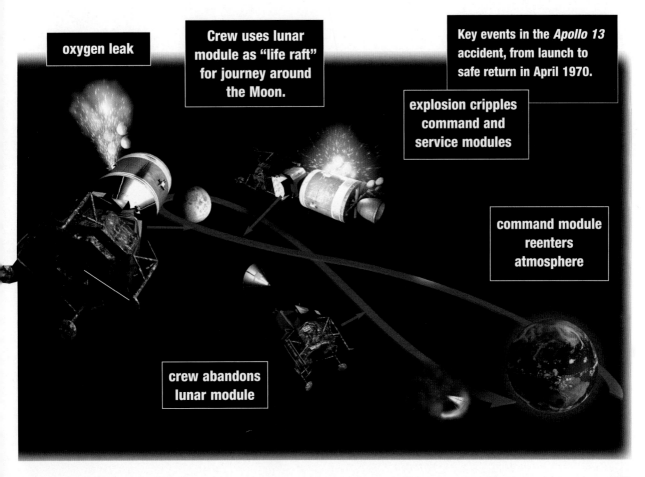

oxygen leak

Crew uses lunar module as "life raft" for journey around the Moon.

Key events in the *Apollo 13* accident, from launch to safe return in April 1970.

explosion cripples command and service modules

command module reenters atmosphere

crew abandons lunar module

# CRISIS ON *MIR*

The Russian space station *Mir*, launched by the former Soviet Union in 1986, had a long and successful life. In the 1990s, as budget cuts hit the Russian space program, *Mir* ran into several problems. In February 1997, an air-supply canister caught fire, filling the station with smoke. The six crew members—four Russians, one American, and one German, could not reach one of the station's two "lifeboat" space capsules. They considered abandoning *Mir* before eventually getting the fire under control.

Four months later, *Mir* had more problems when an unmanned cargo spacecraft rammed the station during a docking test, damaging one of the laboratory modules and a power-generating solar panel.

# THE SHUTTLE DISASTERS

In terms of human life, the most costly space accidents have been the losses of two NASA space shuttles—*Challenger* in January 1986 and *Columbia* in February 2003. *Challenger* was lost when freezing temperatures made one of the rubberized "O-ring" seals on its solid rocket boosters brittle. As the shuttle began its ascent, fuel burning inside the booster rocket formed an intense jet of flame that escaped through a freeze fracture in the seal. The flame burned through one of the struts that attached the booster to the shuttle's main fuel tank, sending the booster crashing into the tank. Less than two minutes into takeoff, an immense explosion resulted, destroying the entire spacecraft and killing all seven astronauts on board. The crew included Christa McAuliffe, who would have been the first teacher in space.

After the accident, NASA identified the problem and discovered many other safety flaws on the shuttle. The entire fleet was grounded for more than two years while changes were made. Even these changes were not enough to prevent another shuttle accident years later.

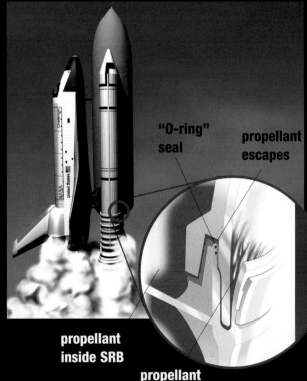

"O-ring" seal

propellant escapes

propellant inside SRB

propellant spontaneously combusts

The cause of the *Challenger* disaster: a brittle "O-ring" seal on one of the solid rocket boosters (SRBs) fractured during launch, burned through the support strut, and crashed the SRB into the external fuel tank.

Throughout the rest of 1997 and 1998, *Mir* experienced more problems, including computer crashes and an air leak. In August 1999, the Russians abandoned *Mir*. Despite several attempts to revive it, the Russian Space Agency eventually decided the station was beyond repair. In 2001, mission controllers commanded it to reenter the atmosphere, breaking up over the South Pacific. Despite its problems, *Mir* was a triumph, operating for ten years beyond its planned five-year life span.

The Russian space station *Mir* hangs above Earth during the last months of its life.

*Columbia* was lost on February 1, 2003 at the end of its mission, breaking up as it reentered Earth's atmosphere over Texas. Once again, all seven astronauts on board were killed, including Ilan Ramon, the first Israeli astronaut. The cause of this tragedy was soon traced back to a problem at launch. Vibrations during liftoff had apparently caused part of the insulation around the main fuel tank to break off and collide with the shuttle's hull, dislodging an area of the shuttle's vital thermal tiling on *Columbia*'s left wing. This led to another long suspension of shuttle flights while more changes were made.

The shuttle finally returned to space for a tense "Return to Flight" mission in August 2005. The crew of *Discovery* carried out a thorough inspection of the shuttle's underside while in space, and discovered that two pieces of packing from between the thermal tiles had come loose. They conducted a unique repair in orbit to remove the pieces of packing that were sticking out and could have caused temperatures to rise during reentry. The discovery that, despite improvements, pieces of the fuel tank insulation had broken away during *Discovery*'s launch has led to more delays in resuming normal shuttle flights.

Wreckage from the space shuttle *Columbia* is laid out in a warehouse according to its location on the shuttle during the accident investigation of 2003.

# GLOSSARY

**Cerenkov effect**: random flashes caused by cosmic rays; seen by astronauts while in space.

**cosmonaut**: an astronaut from the former Soviet Union or present-day Russia.

**deployment**: to put something into service.

**gravity**: a strong, one-way attractive force between two physical bodies; the larger the mass of an object, the greater its gravitational pull.

**impregnate**: to thoroughly saturate a fabric with a liquid.

**infrared**: invisible form of radiation given off by hot objects; also called heat radiation.

**molecule**: the smallest part of a substance, made up of one or a group of atoms, that has all the properties of the whole substance.

**orbit**: the (usually elliptical) path taken by one object around another in space.

**parabola**: a curved bowl shape, such as that of a satellite dish; can also be the shape of a path.

**prototype**: an original example or pattern on which something is modeled.

**radiation**: usually invisible energy transmitted in the form of waves or particles; light waves are the only form of visible radiation.

**reentry**: the act of entering the atmosphere of a planet from the vacuum of space.

**retrorocket**: a rocket engine found on a spacecraft that slows down the craft by creating thrust in the opposite direction.

**satellite**: a natural or artificial object that orbits a planet on a regular path.

**simulate**: to copy or imitate an object or experience; for example, weightlessness.

**space age**: another name for the era of space exploration that began on October 5, 1957, with the launch of Sputnik by the Soviet Union.

**space junk**: debris of all sizes—such as old satellites, hatches blown off space modules, or flecks of paint from spacecraft—orbiting Earth. Space junk threatens spacecraft and creates meteors.

**spectrum**: an ordered arrangement that can be broken down into various components; for example, the colors in white light or the different wavelengths of electromagnetic radiation.

**ultraviolet**: damaging, invisible electromagnetic radiation with wavelengths shorter than visible light but longer than X-rays.

**vacuum**: an area completely empty of matter, even air; space is a vacuum.

**vanes**: the flat surfaces that flare out near the bottom of a rocket to help guide its flight.

**zero gravity**: the state or condition of continuous free fall that feels as if one is free of the pull of gravity—in other words, weightlessness.

# FURTHER INFORMATION

## BOOKS

Cole, Michael D. *Countdown to Space* (series). Enslow (2003).

Collins, Michael. *Flying to the Moon: An Astronaut's Story*. Farrar, Straus and Giroux (1994).

Kerrod, Robin. *Space Pioneers.* The History of Space Exploration (series). World Almanac® Library (2005).

Kraft, Chris. *Flight! My Life in Mission Control*. Penguin (2001).

Kranz, Eugene. *Failure is Not an Option. Mission Control from Mercury to Apollo 13 and Beyond.* Berkeley (2001).

Lovell, Jim. *Lost Moon: The Perilous Voyage of Apollo 13.* Simon & Schuster (1994).

Rau, Dana Meachen. *Space Walks.* Compass Point Books (2004).

Winchester, Jim (Ed.). *Space Missions.* Thunder Bay Press (2006).

Young, Emma. *The Astronaut's Survival Guide*. Puffin (2004).

## WEB SITES

*www.nasa.gov*
Explore NASA's main site to learn about the latest missions and activities.

*www.space.com*
Learn the latest in daily space news.

*www.nineplanets.org*
Take a multimedia tour of the solar system.

*www.spaceday.org/index.html*
Check out NASA's annual Space Day events.

*www.astronautix.com*
Follow links to more information on space exploration.

**Publisher's note to educators and parents:** Our editors have carefully reviewed these Web sites to ensure that they are suitable for children. Many Web sites change frequently, however, and we cannot guarantee that a site's future contents will continue to meet our high standards of quality and educational value. Be advised that children should be closely supervised whenever they access the Internet.

# INDEX